Farid Saadaoui

Energy Management of a Hybrid Energy System (PV/PEMFC/Li-Ion Battery)

Farid Saadaoui

Energy Management of a Hybrid Energy System (PV/PEMFC/Li-Ion Battery)

Energy Management of a Hybrid Energy System (PV/PEMFC/Li-Ion Battery) Based on Hydrogen Minimization Modeled by MER

Noor Publishing

Imprint

Cover image: www.ingimage.com

Publisher:
Noor Publishing
is a trademark of
Dodo Books Indian Ocean Ltd. and OmniScriptum S.R.L publishing group

120 High Road, East Finchley, London, N2 9ED, United Kingdom
Str. Armeneasca 28/1, office 1, Chisinau MD-2012, Republic of Moldova, Europe
Managing Directors: Ieva Konstantinova, Victoria Ursu
info@omniscriptum.com

Printed at: see last page
ISBN: 978-620-0-04866-0

Energy Management of a Hybrid Energy System (PV / PEMFC and Lithium-Ion Battery) Based on Hydrogen Minimization Modeled by Macroscopic Energy Representation

Farid SAADAOUI

1) LIS2T Intelligence laboratory Artificial for on-board systems, advanced telecoms and telemedicine. Electrical Engineering Department, University Mustapha Stambouli, Mascara. Algeria.

2) CAOSEE Research Laboratory Control, Analysis and Optimization of Systems Electro-Energetic systems, University of Tahri Mohamed Bechar, Bp417, Bechar, Algeria.

Email: farid.saadaoui@univ-mascara.dz

Email: fsaadaoui1969@gmail.com

Abstract: This research work is designed for the management of the electric power of an autonomous hybrid system which generally integrates several subsystems, whose main source of production is solar energy (photovoltaic panels) coupled with a hydrogen fuel cell using a storage device (lithium battery).

This energy coupling behavior is used in a wide range of operating conditions ensuring the originality of the exploitation of the energy produced to supply electricity to remote regions and isolated urban regions of southern Algeria, which will be modeled by a recent graphic formalism methodology macroscopic energy representation and controlled by a simple method the maximum control structure that takes into account all the inputs and outputs of the system. This hybrid system is controlled by an energy management strategy by acting on a common continuous bus with variable residential load via a DC/DC converter, allowing control of the amount of energy between the different energy resources to minimize the use of the fuel cell from which to minimize hydrogen consumption. Another is used to maintain the voltage of this bus at its reference via the battery by regulating the bidirectional DC/DC converter.

Keywords: HES, RES, PV, PEMFC, Li-Ion Battery, MER, MCS.

1. Introduction

Electricity grids cannot be sufficient to provide electrical energy to all places, whether they are remote, isolated, mountainous or on islands where connection is prohibitive or in desert environments.

Since the turn of the millennium, the energy sector has undergone two opposing transformations driven by the need to reduce CO_2 emissions while maintaining reliability [1].

On the one hand, the electricity sector, characterized by large centralized fossil power plants, has become increasingly decentralized thanks to the integration of distributed Renewable Energy Sources (RES) [2]. On the other hand, the emergence of these two phenomena represents the new energy transition. In which the development of complementary technologies such as energy storage options, smart grid controls, and demand-side management helps to reliably integrate RES into the existing system while reducing their dependence on energy networks [3].

Multi-energy hybrid systems allow new value propositions to increase the self-sufficiency of electricity networks, flexible sector coupling leading to increased reliability and efficiency, reduced CO2 footprints and improved economy [4][5].

The shortage of power supply has led to the attenuation of economic progress in these areas. However, renewable energy resources are living existing deployed to provide electricity to remote places. Recently, there has been a considerable reduction in the cost of these technologies due to technological advancements nicely government policies; that encourage the use of these technologies [6].

In general, renewable electric power generation systems are very dubious. They are becoming increasingly important in production, where the method of managing the distribution of this off-grid energy is also relevant [7] [8].

It increases the performance of the multi-source system by reducing the hydrogen consumption used by the auxiliary or emergency source; while respecting the demand and the various operating constraints of the energy chain and the system's safety.

The objective of this study is to contribute to the management of hybrid energy production of a multi-physics system composed of a photo-voltaic generator, a fuel cell and a storage device to supply a residential load isolated.

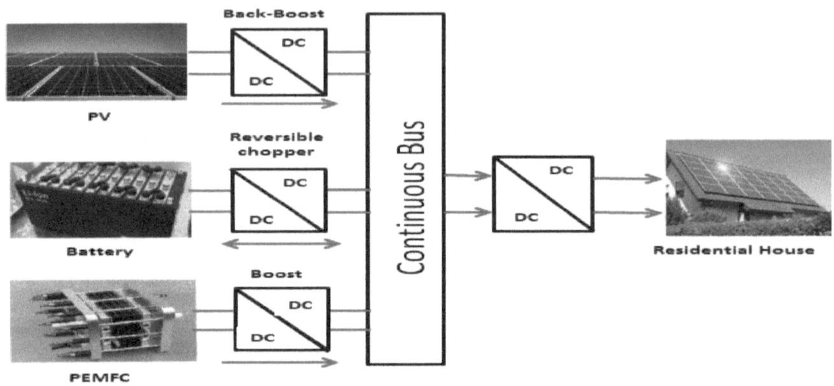

Fig 1: Architecture of the hybrid energy system

The first step is to develop a detailed and in-depth dynamic model under Matlab/Simulink for each system component. A Management strategy has been proposed and applied for each configuration of the system. Or these configurations will be modeled by a recent graphical formalism used for multi-physical systems called the Macroscopic Energy Representation (MER).

The main handicap of Hybrid Energy Systems (HES) is how to use and manage the energy optimally produced by these different resources and how to meet the demand of the load, whatever the climatic variations [9].

Therefore, an efficient management system is necessary to find the best power distribution between different elements constituting the HES.

The randomness, therefore, represents the main drawback of these resources. For this, they are often coupled with other energy sources to constitute HES, to combine the advantages of these resources and minimize their disadvantages to meet the requirements of the load.

The overall control strategy for power management between the different power sources in the system is explained, below. The performances for various operating conditions are evaluated and studied.

The HES can be combined, with other emergency elements, such as wind power or conventional sources, such as the diesel generator. But the latter poses enormous disadvantages (higher cost, conveyance, and pollution). The PEMFC fuel cell is the best element to ensure the load needs throughout the year, which can select from solid oxide fuel cells (SOFC) and molten carbonate fuel cells (MCFC) [10] due to their greater operating flexibility, faster dynamics, and cheaper more efficient [11].

The battery, this storage device, despite its limited lifetime and relatively high cost, is a good element for coupling with the HES. The battery is used for power generation in remote locations to improve and minimize fluctuations in production due to the random nature of these resources, hence the need to underestimate its use and avoid its degradation through deep discharges or overloads.

The main objective of this study is, therefore, the management of the energy supplied by these energy sources to exploit an adequate power distribution according to a sufficient strategy between the different elements constituting the SEH to:

- Ensure the proper functioning of the HES and extend its lifespan;
- Reduce fuel consumption and show the beneficial interest of a hybrid system compared to the conventional energy system;
- Reduce hydrogen consumption for the PEMFC and Reduce the use of storage elements to a minimum;
- Optimize the energy extracted by the main sources of the SEH through the MPPT control;
- Apply this study in the industrial field.

The simplicity and clarity of this methodology on the modeling side by the MER and on the control side by the MCS show the novelty and the impact of this work compared to the currently available published work.

By considering the three models of energy systems by the MER

2. Choice of the MER Modelling Tool

The MER is a tool for the graphical representation of systems. It was introduced in the 2000s to describe complex electromechanical systems. Since then, MER has been developed for complex multi-physical systems (electrical, mechanical, thermal, chemical, fluid, etc.).

It is based, on the principle of action and reaction, to organizing the interactions of different subsystems by their physical causalities (integral causality).

This description highlights the energy properties of a system (accumulation, conversion, and distribution of energy). In addition, Maximum Control Structure (MCS) of a system can always be deduced by the inversion rules of MER. Compared to other graphical representation tools, such as the Bond Graph or the Causal Information Graph (GIC), the MER offers a more global energy view and thus helps to understand the control system. The MER primarily focuses on the energy functions of a system and not on its physical structure [12].

It is a macroscopic extension of the ICG with an explicit symbolization of energy couplings [13]. It allows the definition of a synthetic representation of complex electromechanical systems. This tool makes it possible to describe a system without having an excessively dense graphical reading (as for the GIC). It is based, in an implicit way, on integral causality, which underlies the concept of energy [14].

This concept reveals elements: source, accumulation of energy and conversion without accumulation of energy. These components are all interconnected according to the principle of action and reaction. Finally, the control laws defined by the GIC were extended for the MER.

This made it possible to develop a methodology for the systematic deduction of control structures [15] [16].

2.1 Operating principle of the MER

The MER is made up of an association of objects, linked together by components symbolized by interaction vectors. Figure 2, divides the system into several subsystems. Each subsystem does an action (a1) on the other system and at the same time receives a reaction (r1) from the other system, so there is an exchange of power between the two. The source S1 is an upstream source which produces an action (a1) which is transmitted to the downstream source S2 generating a reaction (r1) [12] [13] [14].

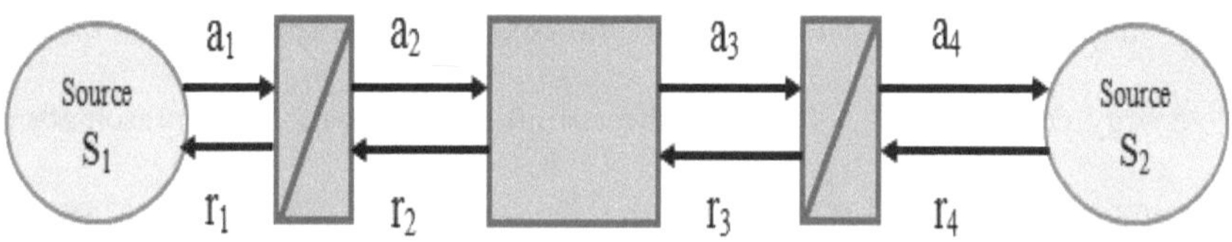

Fig 2: The MER of a conversion chain [12] [13] [14]

2.2 The Maximum Control Structure MCS

The MCS is figuring from the MER according to the principle of inversion: to each element corresponds a control block whose objective is to invert its functionality. It is named MCS because it requires a maximum number of sensors and a maximum number of operations.

There are two types of inversion to ensure the modification of energy:

-Direct inversion for MER conversion elements (without controller)

-Indirect inversion or by servo-control for the storage elements of the MER (with controllers).

Table 1- The basic elements of MER [12] [13] [14] [15][16]			
Macroscopic Energetic Representation MER			
	Action and reaction variables		Energy source (system terminals)
	Energy source (system terminals)		Mono-physical converter (energy conversion)
	Multi-physical converter (energy conversion)		Mono-physical coupling (energy distribution)
	Multi-physical coupling(energy distribution)		

Table 2- The basic elements of MCS [12] [13] [14] [15][16]			
Maximum Control Structure MCS			
	Indirect inversion (closed-loop control)		Direct inversion (open-loop control)
	Direct inversion using a disturbance rejection		Strategy (energy management)
	Coupling inversion (weighting) Model or estimator		Coupling inversion (distribution)
	(any pictogram)		Model or estimator

A detailed description of the elements of the MCS principle and the two types of conversion is given; by the blue parallelograms in Table 2 [12] [13] [14] [15]

[16].

In summary, MER is identified as the most suitable methodology. The simplicity of the adaptation highlights the advantage of using a modular approach to modeling.

The MCS is achieved by block-by-block inversion, where the choice of controllers and their settings are left to the developer. The control application shows that by using MER, it is possible to derive the control structure from the model of a complex multi-domain system without the need for the transfer function of the whole system.

3. Modeling of the hybrid system by the MER

3.1Modeling of the PEMFC

The PEMFC (Proton Electrolyte Membrane Fuel Cell) as an auxiliary emergency source and non-reversible hydrogen destocking chain; or it can be reversible if it combines the hydrogen storage chain via the electrolyser (in this study we will assume that the hydrogen is stored in hydrogen cylinders). The heart of the PEMFC consists of [17]:

-Anode: an emitter of electrons in the reaction is: $2H_2 \rightarrow 4H^+ + 4\,e^-$ (1)

-Cathode: an electron receptor in the reaction is: $O_2 + 4\,H^+ + 4\,e^- \rightarrow 2\,H_2O$ (2)

-The support membrane (electrolyte) separates the anode from the cathode in the reaction is:

$$H_2 + \frac{1}{2}O_2 \rightarrow H_2O + heat + electrical\ energy \quad (3)$$

The voltage of a cell is given according to Ohm's law:

$$V_{cell} = 1.29 - 0.85\times10^{-3}\times(T-298.15) + 4.31\times10^{-5}\times T\times\left[ln(P_{H_2}) + \frac{1}{2}ln(P_{O_2})\right] - \left[\xi_1 + \xi_2 T + \xi_3 T ln(C_{O_2}) + \xi_4 ln(I_{stack})\right] - Bln\left(1-\frac{J}{J\max}\right) - I_{stack}\left(\frac{t_m}{\sigma}+Rc\right) \quad (4)$$

P_{H_2} : is the partial pressures of hydrogen (atm), P_{O_2} : is the partial pressures of oxygen (atm) and T the fuel cell temperature (°K).

Where $\xi_1 = -0.948$; $\xi_2 = 0.00286 + 0.0002ln(\text{A}) + \left(4.3\times10^{-5}\right)ln(C_{H_2})$; $\xi_3 = 7.6\times10^{-5}$ and $\xi_4 = -1.93\times10^{-4}$: are experimentally defined parametric coefficients.

$B = \frac{RT}{nF} = 0.016V$: is a parametric coefficient, which depends on the cell, t_m is the operation state and J represents the actual current density of the cell (A/cm^2).

J_{max} = 1.5A/cm^2 is the current density limit and t_m: is the membrane thickness.

The oxygen concentration given by [18] in (mol/cm^3):

$$Co_2 = \frac{Po_2}{5.08\times10^6\times e^{(-498/T)}} \quad (5)$$

σ_m: Is the specific conductivity of the membrane proton [19]:

$$\sigma_m = (0.00519\lambda - 0.00324)\exp\left(1268\left(\frac{1}{303}-\frac{1}{T}\right)\right) \quad (6)$$

λ is the water content of the membrane given by the following relation [20].

$$\lambda = \begin{cases} 0.043 + 17.81RH - 39.85(RH)^2 + 36(\text{RH})^3, & 0 \le RH \le 1 \\ 14 + 1.4(\text{RH}-1) & 1 \le RH \le 3 \end{cases} \quad \text{RH en \%} \quad (7)$$

The partial pressures of the three inlet gases to the PEMFC cell are given by [20]:

$$\frac{d}{dt}(P_{H2})=\frac{RT}{V_{an}}(q_{H_2in}-q_{H_2out}-q^{r}_{H_2}) \tag{8}$$

$$\frac{d}{dt}(P_{O2})=\frac{RT}{V_{an}}(q_{O_{2in}}-q_{O_2out}-q^{r}_{O_2}) \tag{9}$$

$$\frac{d}{dt}(P_{H_2O})=\frac{RT}{V_{an}}(q_{H_2Oin}-q_{H_2Oout}-q^{r}_{H_2O}) \tag{10}$$

P_{H_2inc}, P_{O_2inc} and P_{H_2Oinc}: the controlled partial pressures of each gas inside the fuel cell (atm).

V_{an} : is the volume of the anode (l), q_{H_2}, q_{O_2} : is the flow rate inlet of each gas and q_{H_2O} : is the water inlet flow rate by air (kmol/s).

q_{H_2out}, q_{O_2out} : are the outlet flow rates of each gas and $q_{H_2O_{out}}$: water vapor on (kmol/s).

q_{H_2in} and q_{O_2in}: the inlet flow rates of hydrogen and oxygen of the cathode and anode;

q^{r}_{H2}, q^{r}_{O2} and q^{r}_{H2O}: usage, production also reaction of the gases and water.

The fuel cell model previously depicted by the MER is clearly shown in figure 3.

$$q^{r}_{H_2}=2q^{r}_{O_2}=q^{r}_{H_2O}=\frac{N_0 I_{stack}}{2F}=2K_r I_{stack} \tag{11}$$

$$q_{H_2out}=K_{H_2}P_{H_2};q_{O_2out}=K_{O_2}P_{O_2};q_{H_2Oout}=K_{H_2O}P_{H_2O} \tag{12}$$

The formula of relative humidity of a fuel cell is given by:

$$RH=\frac{P_{H_2O}}{P_{sat}(Tair)} \tag{13}$$

P_{H_2O} : The partial pressure of water vapor and P_{sat} : The vapor saturation pressure.

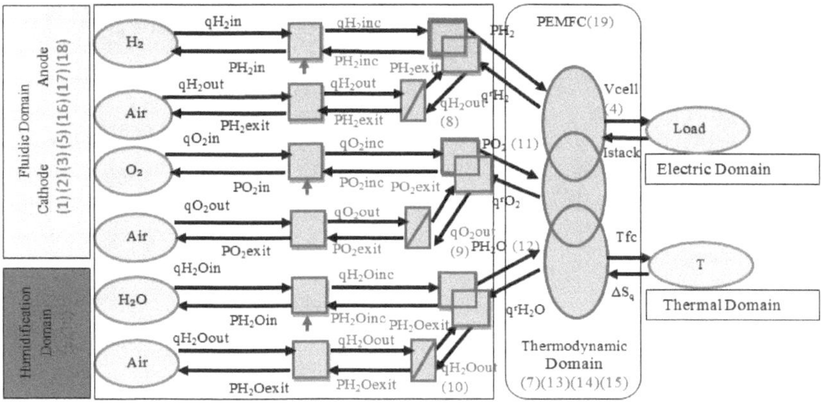

Fig 3: The proposed representation of the PEMFC by the MER

$$P_{sat} = 10^5 \exp\left(13.7 - \left(\frac{5120}{T_{air} + 273.15}\right)\right) \tag{14}$$

$$P_{H_2O} = P_{exit}\left(\frac{0.420 + \lambda_a \psi}{\lambda(1+\psi) + 0.210}\right) \tag{15}$$

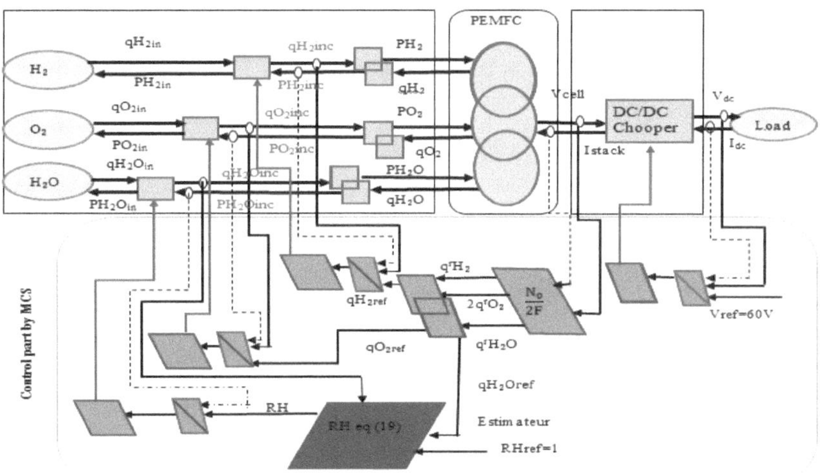

Fig 4: The Proposed Synoptic Control Strategy of PEMFC using MCS

Table 3- Model Parameters [21]

PARAMETERS	VALUES
Faradays Constant, F	9684600 C/Kmol
Universal gas Constant, R	8314.47 j/Kmol.K
Number of cells, N_0	100
Hydrogen valve constant, KH2	$4.22*10^{-5}$ Kmol/S.A
Oxygen valve constant, KO2	$2.11*10^{-5}$ Kmol/ S.A
Stack internal resistance, Rc	0.00003 Ω
Constant, Kr	$0.996*10^{-6}$ Kmol/ S.A
Surface, A	50.6 cm^2
Membrane thickness, tm	178 μm
Atmospheric fuel cell λ_a	2
Current stack, Istack	60
Adjustable parameter; ψ	23

P_{exit} : is the pressure at the output of the stack and ψ the air stoichiometry given by:

$$\psi = \left(\frac{q_{H_2Oin}}{q_{O_2in} + q_{rest}} \right) \tag{16}$$

$$q_{O_2in} = \frac{\lambda_a . I_{stack}}{4.F} \tag{17}$$

and the molar flow of oxygen remaining $q_{rest} = 3.76.q_{O_2in}$ (18)

So the relative humidity becomes (%):

$$RH = \frac{P_{H_2O}}{P_{sat}(T)} = \frac{P_{exit}\left(\frac{0.420 + \lambda_a \psi}{\lambda(1+\psi) + 0.210} \right)}{10^5 \exp\left(13.7 - \left(\frac{5120}{T_{air} + 273.15} \right) \right)} \tag{19}$$

3.2Modeling of the PV generator

a) Modeling of the Photovoltaic Cell

The photovoltaic panel as a non-reversible main source consists of several silicon cells connected in series and parallel. They absorb solar energy and transform it into electric current. This transformation is carried out by converters to regulate the output voltage and current of this solar cell, figure 5.

$$I_{Ph} = I_D + I + I_{Rp} \quad (20)$$

$$I_D = I_s\left[\exp\left(\frac{q(V + IR_S)}{nKT}\right) - 1\right] \quad (21)$$

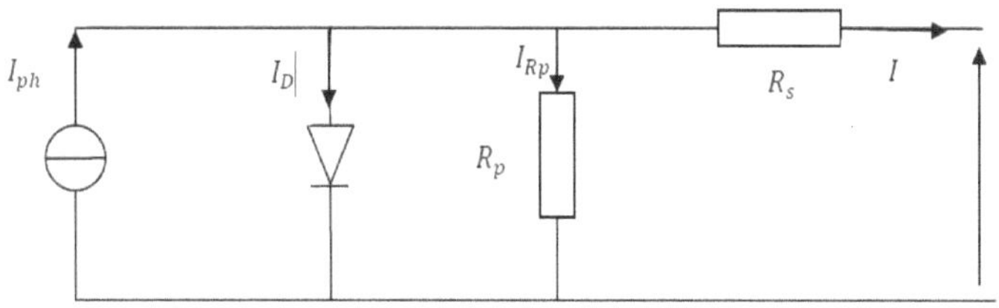

Fig 5: Equivalent circuit of solar cell

Is : is the saturation current of the diode

$$I_s = K_1 T^3 \exp\left(\frac{-E_g}{KT}\right) \quad (22)$$

Where T: is the effective temperature of the cell in °K

$$V_t = \frac{KT}{q} \quad (23)$$

$$I_{Rp} = \frac{V + IR_S}{R_P} \quad (24)$$

From equation (20) we obtain the expression of the characteristic I-V of the chosen model.

$$I = I_{Ph} - I_s\left[\exp\left(\frac{V + IR_S}{nV_t}\right) - 1\right] - \frac{V + IR_s}{Rp} \quad (25)$$

$$I_{Ph} = (I_{sc} + K_i dT)\frac{G}{G_n} \quad (26)$$

G : Solar radiation and G_n : Nominal solar radiation

$$I_s = \frac{I_{sc} + K_i dT}{\exp\left(\frac{V_{0C} + K_V dT}{V_t a}\right) - 1} \quad (27)$$

$$I = N_{PP} I_{Ph} - N_{PP} I_s \left[\exp\left(\frac{V + (R_S N_{SS}/N_{PP})I}{V_t a}\right) - 1\right] - \frac{V + (R_S N_{SS}/N_{PP})I}{R_P N_{SS}/N_{PP}} \quad (28)$$

Table 4- The Electrical Characteristics of Photovoltaic Module [22]

Parameters	Values
Gr: Solar radiation reference	1000 (W/m^2)
Rs: Equivalent series resistance of the resistive contacts	0.5 Ohms
Rsh: Shunt resistance representing the leaking currents	500 Ohms
q: Electron charge	$1.602*10^{-19}$ C
K: Boltzmann constant	$1.38*10^{-23}$ J/°K
T : is the effective temperature of the cell	°K
Tinit: Initial temperature of the PV Cell	298 °K
Isc: Short-circuit current	3.33 A
Voc : Initial open-circuit voltage	21.3 V
Ki : Temperature versus current coefficient	0.0025A/°C
Eg : Gap energy silicium	1.12 eV
N_{PP} : is the number of photovoltaic modules in parallel	1
N_{SS} : is the number of photovoltaic modules in serie	36

The choice of a PV model generally depends on the intended use of the model and the information available to determine its parameters. Each model should be built on classical principles [23] Table 4.

-PV cells have the same characteristics,

-PV depend on sunlight, solar incidence and temperature,

-PV does not have partial shadows on their surfaces.

To improve the electrical management of PV systems, it is essential to model and analyze these specific intrinsic characteristics [24]. As shown before in the PEMFC, the Maximum MCS is deduced from the modeling of the PV system by the MER, using inversion rules cited in Table 2 as shown in figure 6. This control structure is taken out to supervise the operating point of the source and the load and also to control the stability of the DC bus, as well as all the components of the system.

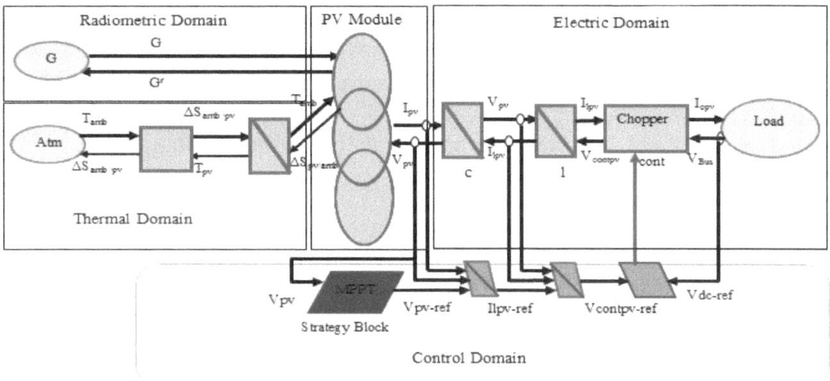

Fig 6: The macroscopic energy representation of a PV

The MCS is composed of two indirect reversal controllers, a strategic block of MPPT as well as a direct reversal whose entry is a distribution criterion α.

From the radiation, the module temperature at the estimated or measured wind speed, the open-circuit voltage is calculated and the MPPT is deduced. Using a PI voltage controller, a reference current is obtained, and a PI current controller deduces the reference voltage at the input of the chopper which is divided by the bus voltage to generate the conversion function for the chopper control [25].

3.3Lithium-Ion Battery Modeling

By convention, the battery is seen as a voltage generator. The charge current is negative, and the discharge current is positive. For this model, the influence of temperature is neglected, the capacitance model is approximated by capacitor C_b and the voltage model is obtained by applying the mesh law [26]. Figure 7.

$$V_{bat} = E - R_0 i_{bat} + V_{C_b} \tag{29}$$

Where i_{bat}: is the battery current, E: is the open circuit voltage of the battery. Resistance R_0 represents the internal resistance of the battery, while the ideal voltage source is in series with the capacitor C_b models the linear part on the charge and discharge curves.

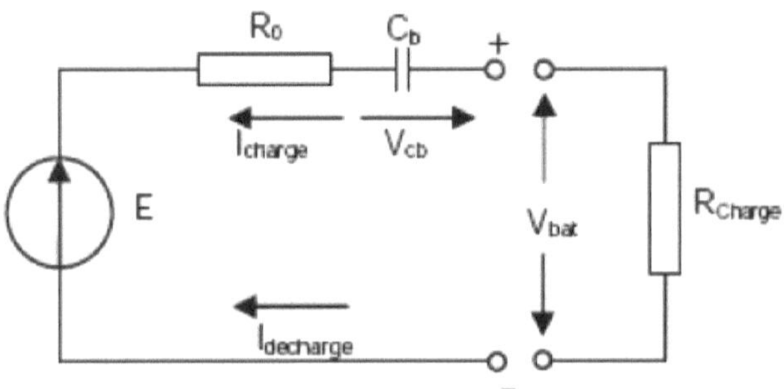

Fig 7: Simplified electric model of a battery

a) Battery discharge model (i *> 0)

$$f1(it,i*,i) = E0 - K\cdot(Q/(Q-it))\cdot i* - K\cdot(Q/(Q-it))\cdot it + A\cdot exp(-B\cdot it) \tag{30}$$

b) Battery Charging Model (i* < 0)

$$f2(it,i*,i) = E0 - K\cdot(Q/(it+0,1\cdot Q))\cdot i* - K\cdot(Q/(Q-it))\cdot it + A\cdot exp(-B\cdot it) \tag{31}$$

With the characteristics given by the manufacturers:

Ebatt = nonlinear voltage in (V), E0 = constant voltage in (V),

K = Polarization constant in (Ah−1) or polarization resistance (Ohms), i* = dynamic low frequency current (A), i = battery current (A),

it = Capacity extracted (Ah), Q = Maximum battery capacity (Ah),

A = Exponential voltage (V), B = Exponential capacity (Ah−1)

The battery can be seen as a reservoir of energy, the quantity of which is constantly changing. Consequently, its state of charge is identified as the capacity contained in this battery. It is affected by the operating conditions (current, temperature, etc.).

The SOC unit is 100% or 1 for a fully charged battery and 0% or 0 for a so-called empty battery. *SOC(t)* is the state of charge of the battery expressed by the following relation with Q is the nominal capacity of the battery [26] [27].

$$SOC(t) = 100\left(1 - \frac{1}{Q}\int idt\right) \tag{32}$$

In order to avoid degradation of the battery and prolong its life, their state of charge must be maintained within a certain interval defined as follows [26] [27]:

$$SOC(t)_{min} \leq SOC(t) \leq SOC(t)_{max} \tag{33}$$

$SOC(t)_{max}$: Maximum battery charge state and $SOC(t)_{min}$: Minimum battery charge state.

The Lithium-Ion battery as an auxiliary source, it is used both as a source in the case of a production deficit and as a burden in the case of overproduction.

Table 5- The characteristic parameters of a Lithium-Ion battery [26] [27]	
Parameters	Values
Vn : Nominal voltage	14 V
Rs: Nominal resistance	4 Ah
SOC : Initial State of charge	70%
tr : Battery response time	30 s
Cm : Maximum capacity	4 Ah
Vf : Fully charged voltage	16.2958 V
In : Nominal discharge current	1.7391 A
Ri : Internal resistance	0.035 Ohms
Cn : Rated voltage capacity	3.6174 Ah
Exponential zone [Voltage Capacity]	[15. 1254 Ah 0.196522 Ah]

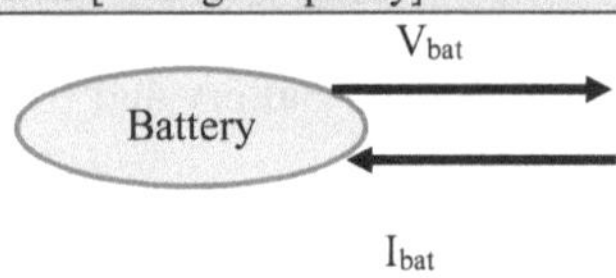

Fig 8: Macroscopic Energy Representation of Li-Ion Battery

4. Hybrid System Components Sizing

4.1. PV generator system sizing

In this study we took the month of January 2019 to determine the lowest average global monthly and daily radiation and temperature values for the year, especially for the town of Bechar (figures 9 and 10).

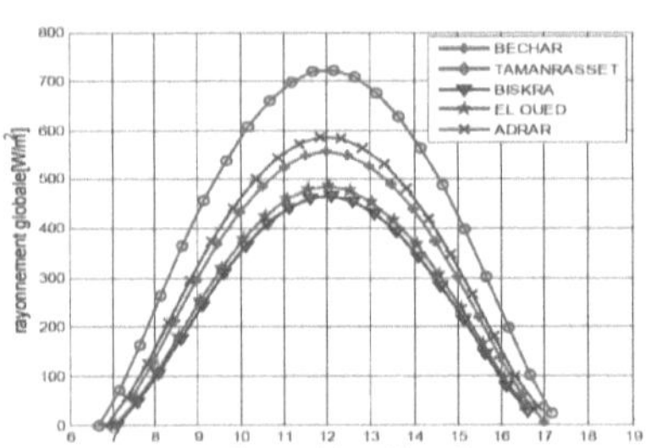

Fig 9: Radiation variation schema for Algerian southern measured by the weather services on 06 January 2019

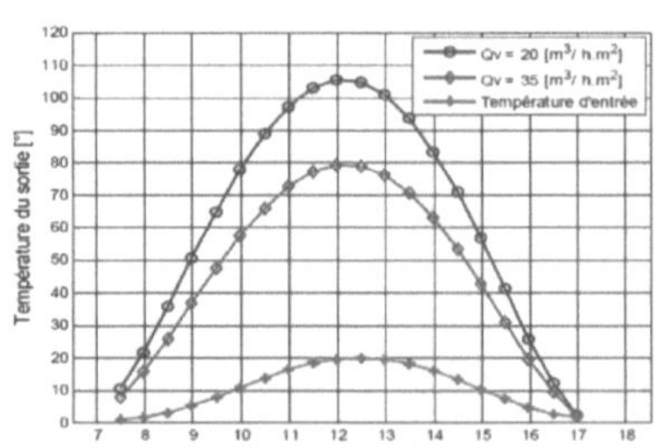

Fig 10: Temperature variation schema for Algerian southern measured by the weather services on 06 January 2019.

a) The consumption power of the load

The above table 6 which is presented in [28] is taken as an example reference in the Bechar region in January 2019. We can notice that the daily consumption energy E=2160Wh/d, this consumption is tolerable more or less. The loss of energy (dissipated) is estimated at 15% of the energy required.

ie: Edissipated (Wh/d) =E*0.15= 2160*0.15=324 Wh/d

The total energy required E =2160 + 324 = 2484Wh/d we take E= 2500 Wh/d domestic consumption varies from season to season; it is higher in the summer than in the winter.

Table 6-Consumption profile of a typical house in Saoura Bechar [28]

Designation	Number	Power(W)	Duration of use(h)	Energy consumed (Wh/j)
Neon Lamp	6	30	18 :00 to 00 :00 (5h)	640
Television	1	50	18 :00 to 00 :00 (6h)	250
Fridge	1	75	07:00 to 20 :00 (13h)	920
Ironer	1	30	19 :00 to 20 :00 (1h)	30
Fan	1	80	14 :00 to 17 :00 (3h)	320
Total		265		2160

-Daily housing requirement

$$B_d = \sum_{1}^{n} P_{load}\, Dt \qquad (34)$$

Where B_d: The daily electricity consumption of a habitable house, Dt: operating time of the consumable device and P_{load} : Power of the device consumable energy.

For the case where there is a load, the current is negative, for example (the regenerative braking of a house elevator or charging of a hybrid vehicle that recovers energy from the battery).

In this context, the load, the contained bus, and the converters are not modeled.

-Daily power of the PV panel

$$P_{pv/d} = P_{load} N_e \tag{35}$$

$P_{pv/d}$: daily power of the module, P_c: peak power of the module.

$N_e = \frac{G_s}{1000}$: Number of equivalent hours.

With G_s: the monthly average solar radiation value of the most unfavourable month of radiation, 1000: solar radiation value under the conditions STC en (W/m^2).

In photovoltaic installations, the losses are approximated at 20%.

$$P_{pv/dactual} = P_{pv/d}(1-0,2) \tag{36}$$

b) The PV field sizing

The following formula shows the size of the photovoltaic field necessary to produce the power of the load [28] [29]:

$$S_{pv} = \frac{E_d}{G\eta_{pv} f_T \eta_{bat}} \tag{37}$$

With:

- S_{pv} : this is the active area of the photovoltaic field in (m2).
- E_d : this is the daily energy consumption in (KWh).
- G : it is the solar radiation incident on the surface in (w/m2).
- η_{pv} : this is the efficiency of the photovoltaic module in (%).
- f_T : this is the temperature correction factor in (%).
- η_{bat} : this is the efficiency of the batteries used in (%).

[28] [29] [30] gives the output of the panel:

$$\eta_{pv} = \eta_0 (1 - f_T (T_{cm} - T_r)) \tag{38}$$

Or η_0: is the efficiency return that reaches 25% for $G=1000\ W/\ m^2$

T_r : reference ambient temperature = 25° C and T_{cm} : internal cell temperature

$$T_{cm} = T_a + \frac{G}{800}(N_{oct} - 20) \tag{39}$$

N_{oct}: represents the nominal temperature of the cell measured at $G= 800\ W\ /\ m^2$, at an ambient temperature of 20 °C and a wind speed of 1 m / s, it is between 40 °C and 70 °C.

G: global radiation of the module and T_a: the ambient temperature of the site.

For a minimum average monthly radiation for the month of January which tends towards $3.16\ KWh/m^2$ and with an energy consumption $Ed=2500Wh/d$, the area of the photovoltaic modules for energy is given by:

$$S_{pv} = \frac{2500}{1000 \times 0.22 \times 0.8 \times 0.85} = \frac{2500}{1.5} \approx 16.8m^2$$

c) Peak power of the PV generator

The peak power of the generator under radiation 1000 W/m2 given by the relation [28] [29] [30]:

$$P_c = S_{pv}\eta_{pv}G \quad (40)$$

That is: $P_c=16.8*0.22*1000=3696W$

d) Calculation of the number of modules

The number of Solarex MSX 60 modules with 60 W peak power that can be used:

$$N_{pv} = \frac{3696}{60} = 61.6\,\text{modules} \approx 62\,\text{modules} = 65 \text{ modules} \quad (41)$$

So we have 65 modules for a photovoltaic field area of 16.8m2.

Parallel modules: 3696W / 220V = 16.8A and 16.8A / 3.5A = 4.8 or 5modules

Modules in series: 220V / 16.8A = 13 modules

4.2.Lithium-Ion Battery sizing

a) Battery capacity

In order to ensure the continuity of electricity production during darkness, at night, when there is no light and during climatic fluctuations, the storage of electricity is essential.

The battery capacity depends on several parameters such as the number of days of autonomy (N), the battery efficiency (ηbat), the maximum depth of discharge (D), the energy consumed (E_d) and the battery voltage (V_{bat}) [28] [29][30]:

$$C_{bat} = \frac{E_d N}{\eta_{bat} D V_{bat}} \quad (42)$$

We take an autonomy of N = 1 days.

D: Maximum depth of discharge D = 70% and the battery voltage of V_{bat} = 24 V

There $C_{bat} = \frac{2500 \times 1}{0.9 \times 0.7 \times 24} = \frac{2500}{15.12} \approx 150Ah$

4.3.PEMFC Sizing

To determine the number and area of cells that make up a PEMFC cell, it is necessary to size it taking into account the nominal power of the FC and the current density that we want to have [31] [32] [33].

a) Number of PEMFC cells

The gross electrical power of the stack in (W) is calculated by the following relation:

$$P_{fc} = NV_{cell}JA \quad (43)$$

N: Number of cells in the stack, *Vcell:* Voltage of cell (V),

J: Current density (A / cm2), *A:* Active cell surface (cm2).

Vcell and *J* are linked by the voltage/current characteristic of the PEMFC, so there remain three independent variables to adjust in order to obtain the desired power. The constraints related to the application will allow us to make the necessary choices. It is interesting to have the highest voltage and therefore the current. It is interesting to have the highest voltage and therefore the lowest possible current, because this limits the joule losses in the cell. To satisfy this constraint, it is possible to:

• Increase the number of cells to the current technological limit.

• Increase the voltage per cell, by decreasing the current density, therefore by increasing the cell surface.

This choice is also in line with the best performance of the PEMFC. It is interesting to have high power densities (to reduce the size and cost per kW of the fuel cell). So implies a small cell area and higher current densities. This constraint opposes the goal of having a high voltage per cell. Too high current density also leads to low efficiency and, therefore, high thermal power to be removed, which will cause problems for the system.

A reasonable limit for the current density is 0.6 A/cm², which corresponds to a voltage of $Vcell = 0.64V$, but this is an arbitrary value that should be validated by a technical-economic study [31] [32] [33].

The PEMFC is supplied; with a 230V household voltage. Or the number of cells will be:

$$N_{cell} = \frac{U}{V_{cell}} \tag{44}$$

$$N_{cell} = \frac{230}{0.7} \approx 330$$

b) The surface of PEMFC cells

It is now possible to determine the other parameters of the stack:

$$I_{max} = \frac{P_{max}}{U} \tag{45}$$

$$A = \frac{I_{max}}{J} \tag{46}$$

A: this is the total area of PEMFC cells

In our application we consider that the maximum useful power delivered by the PV field is 3696W.

I_{max} : Maximum current corresponding to $P_{utile\ max}$, so : $I_{\max} = \frac{3696}{220} \approx 19A$. For safety reasons we consider that $I_{max}=30A$ therefore $A = \frac{30}{0,6} = 50cm^2$, $A=50\ cm^2$.

According to relation (43) the PEMFC can reach a maximum power:

$P_{pile\ max}$ =330*0.7*0.6*50= 6930W.

4.4. Hybrid system by MER

The overall system connected to a residential load via a DC/DC continuous bus, shown in Figure 11, is represented by the MER as a continuation.

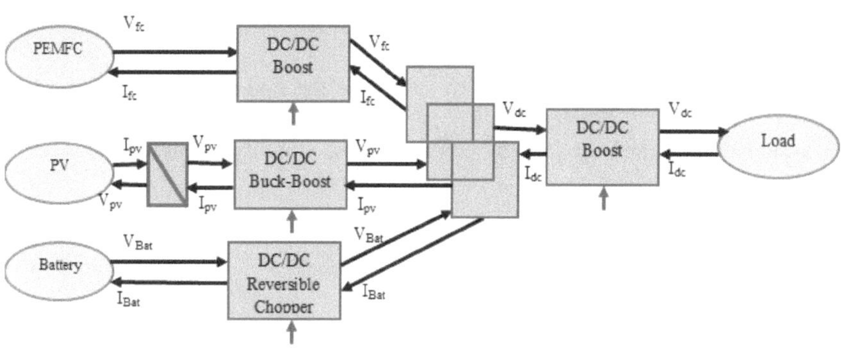

Fig 11: Architecture of the HES by MER

The MER of the hybrid system; is realised in two different steps. The first step is to define the MER of each component. The second step is to represent the MER of the overall system by assembling the MERs of all elements.

This representation; is characterized by the pooling of several physical sources, which can couple in the conversion chain. The implementation of such a system requires specific power management of these couplings.

The simplicity of this formalism is singular by the MCS, which highlights the problems posed by these different couplings. Or the controlled parts (which have a small red arrow at the bottom of the converters in figure 11) of the system can be easily detected.

4.5.Design of the HES control based on the MCS

The MCS uses inversion rules as shown in figure 12, where the MCS is composed of four direct inversion controllers, three indirect inversion controllers, and a block of energy management strategy with a load considered to be controlled.

The objective of the MCS is to supervise the operating point of the hybrid system sources and the load by controlling the stability of each energy element, which leads to the stability of the DC bus.

4.5.1. Reversal principle for the control

The static converter is a dual modulator (modulated current and modulated voltage), and the inverting principle is applied to the power chain starting from the converter of the first source to be controlled (principal PV source) to the last converter (of the PEMFC). Either the control input introduced by the power converter uses to impose on the power chain. It is the voltage or the current across a voltage or current source.

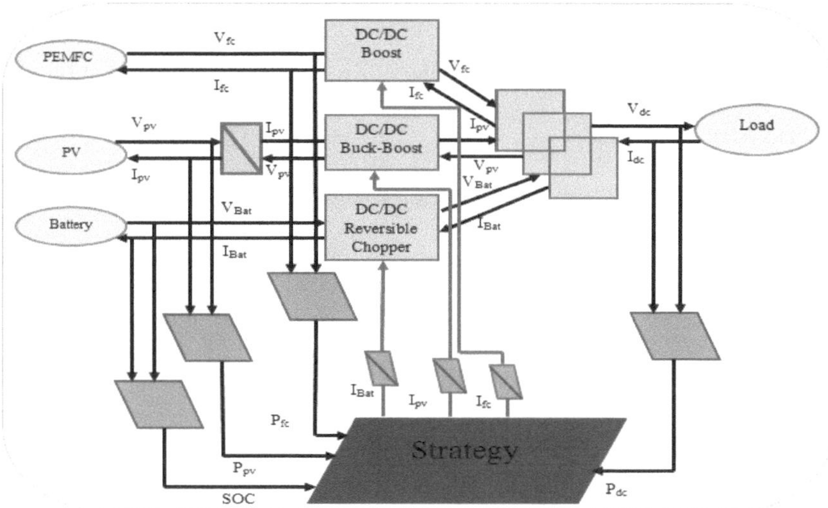

Fig 12: Proposed Control Strategy of HES by MCS

Any production unit must have an energy management supervision that takes into account parameters characterising the proposed multi-source hybrid system (climatic conditions, number of sources used and storage devices, etc.). As a result, a control strategy was developed.

The MER of a hybrid system is obtained by interconnecting the different models of the components and makes it possible to offer each architecture a synthetic representation while respecting the physical causality of the system.

MER makes it possible to define a synthetic representation based on a functional description. From this, we can deduce the MCS according to the following procedure:

1-determination of the MER and the adjustment inputs that allow acting on the system (such as the switching order of static converters)

2-determination of the variables of the system that one wishes to control

3-determination of the adjustment chains that connect the inputs and outputs of the system

4-determination of the MCS control chain, which includes a maximum of control and measurement operations (any quantity is considered measurable). This methodology corresponds to local control; in other words, energy management is as close as possible to the power components.

5-determination of the supervision strategy necessary to control the entire system

This level makes it possible to link the references and the energy criteria, according to a well-defined specification.

The command blocks represent only one manipulation of information. They are all represented by blue parallelograms (homogeneity of the command). The variables exchanged are represented by continuous lines (essential information) or broken lines (optional information).

4.5.2. Objectives of the control part

One of the advantages of MER is that it highlights the degrees of freedom to properly manage energy flows within a system. The objective of this methodology will be to apply the inversion rules of MER to deduce a systematic control for the three architectures studied.

The control inputs are the places; where the control part can act on the system. They are indicated by red arrows in figure 12. In this case, the system is controllable via the auxiliaries.

The first step of the methodology consists in determining the adjustment variables, control objectives and system constraints. In this case, the main is to control PEMFC voltage, PV current and battery voltage.

The second:

-The PV current is controlled through a Buck-Boost converter via a direct inverting blue parallelogram.

-The battery current is controlled by a Reversible-chopper via a direct-inverting blue parallelogram.

-The FC current can be controlled with a power electronics converter placed; between the FC and the electrical load. Generally, this additional degree of freedom used to manage the power supplied by the PEMFC or to limit current dynamics, which could lead to premature ageing of the PEMFC.

Another among the advantages of MER formalism is that it makes it easy to highlight the blocks of strategies (dark blue pictogram). The control part of this method is founded on this strategic block capable of managing the powers of the PEMFC, the PV, the battery and the residential load according to the scenarios proposed in figure 14.

The basic principle of the MCS consists of a step-by-step inversion of the MER following the chains of control. The conversion and coupling elements are directly inverted, while the accumulation elements are inverted; via a controller. Indeed, the direct inversion of an accumulation block is not possible according to the integral causality rule.

The MCS is a theoretical control structure. Therefore all the variables are considered measurable.

5. **Simulation and Results**

5.1. Different operating modes of the HES

The multi-source system consists of a 2.5KW power GPV, a PEMFC system with a nominal power of 2.5kW and a 150 Ah battery bank. The HES is designed to supply a residential 220V load. Before proposing the philosophy of supervision of management and exploitation of the electrical energy produced, we consider that [34] [35]:

- The PV system is considered as the main source.
- The fuel cell is considered a backup source.
- The battery is used both as a source in the case of a production deficit and as a load in the case of overproduction.

According to figure 15 of the management control strategy of the HES and the figure 14, we can distinguish during power exchange between the load and the HES three operational modes can appear as follows [36]:

a) Normal mode

In this mode, the load power is positive and less than the maximum power of the main source (photovoltaic). Interface converter control allows storage devices to be charged or unloaded so that their voltage is equal to their reference. This is part 1 or t∈ [0h, 8h], solar radiation in winter is zero (G = 0 therefore Ppv = 0). The load is supplied from the batteries.

b) Recovery or compensation mode

In this mode, the load power is negative and the power must therefore be able to be absorbed by the multi-source system. It's here clearing part or t ∈ [8h, 9h]. In winter from 8 a.m., radiation begins to appear and Ppv> 0 but the photovoltaic power is insufficient to satisfy the load, which calls for the auxiliary sources of batteries and PEMFC.

c) Overload mode

In this mode, the power absorbed by the load is greater than the power of the main source, the radiation is very large. Control of the interface converter must therefore make it possible to transfer power from the storage devices to the load.

This is the part where t ∈ [9h, 18h], the preferred mode operation.

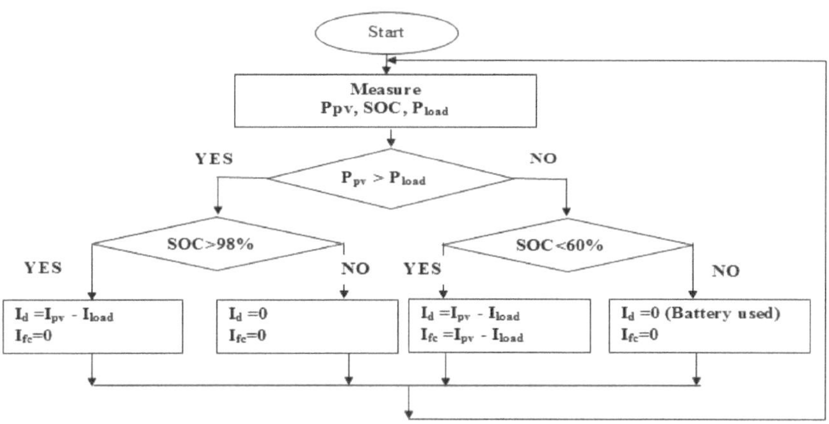

Fig 13: Proposed HES management algorithm

Table 7- HES power source states

HES Power Source Status		ON	**OFF**
Main Source	PV Sytem	Active solar radiation $0< G \leq 1000W/m^2$	in the dark at night or on a cloudy day and in case of PV failure
Auxiliary Sources	PEMFC Source Relief	When the PV is not working and $SOC_{min} < 60\%$	During normal operation of the main source (PV).
	Lithium-Ion Battery	During the inability of the PV 1. In the event of PV failure. 2. The climatic conditions are weak. 3. The demand of the load is greater than the production. When : $60\% \leq SOC \leq 98\%$ 4. In case of excess energy, with SOC <60%.	The power of the load equal to or less than the output power : $\mathbf{Pch \leq Ppv}$
Load	Load DC/DC	The load is still connected.	

Therefore, to simplify the study, we assume that each component of the SEH has two states: ON or OFF. For example, on a summer day, solar radiation is stronger during the day. Table 7 describes the situation of the environment with hybridisation and the state of each component of the SEH. Indeed, depending on the available energy sources used, the weather conditions, the energy consumption and the state of charge of the batteries, several available operating scenarios can be discussed and are detailed in Table 7, following the above-mentioned algorithm [37][38].

5.2.DC bus voltage control

The components of the hybrid system are interconnected via a DC bus whose voltage V is set at 220 V or the voltage of the DC bus will be checked by the battery or by the PEMFC.

5.3. Control the V_{dc} charge voltage by the Battery

In case the PV is unable to meet the load requirements and the DC bus voltage becomes lower than the reference value so if the SOC is at its maximum value, in this case the battery is in operating mode to support the shortage of lost energy.

Conversely, when there is an excess of energy, the DC bus voltage becomes greater than the reference value, the battery is in a state of charge from this excess.

In this case therefore, the voltage of the DC bus V_{dc} is regulated by the battery, via a current reversible chopper through a conventional PI (Proportional-Integral) regulator.

5.4. Control the Vdc charge voltage through the PEMFC.

When the Lithium-Ion battery is unable to control the DC bus voltage $\mathbf{V_{dc}}$, i.e. the SOC is minimal, which causes the intervention of the boost converter that is located between the PEMFC and the bus continuous by a conventional PI corrector to ensure this regulation.

5.5. Management Strategy of HES

The primary goal of this study is to optimally define an energy management strategy for the HES (PV/ PEMFC/Li-Ion battery).

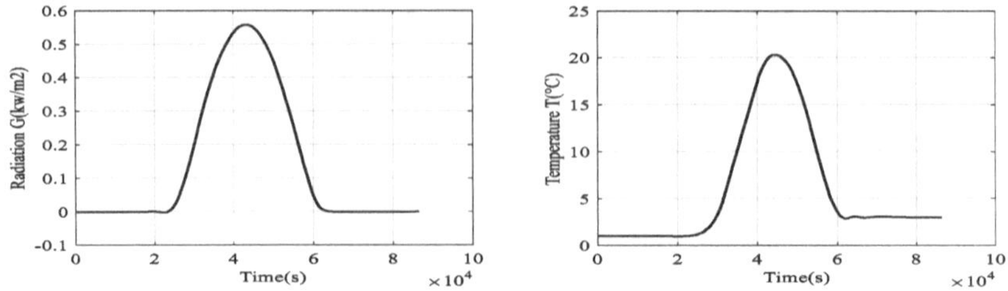

Fig 14: Diagram of variation of radiation and temperature in the city of Bechar, January 6, 2019, lasting 24 hours

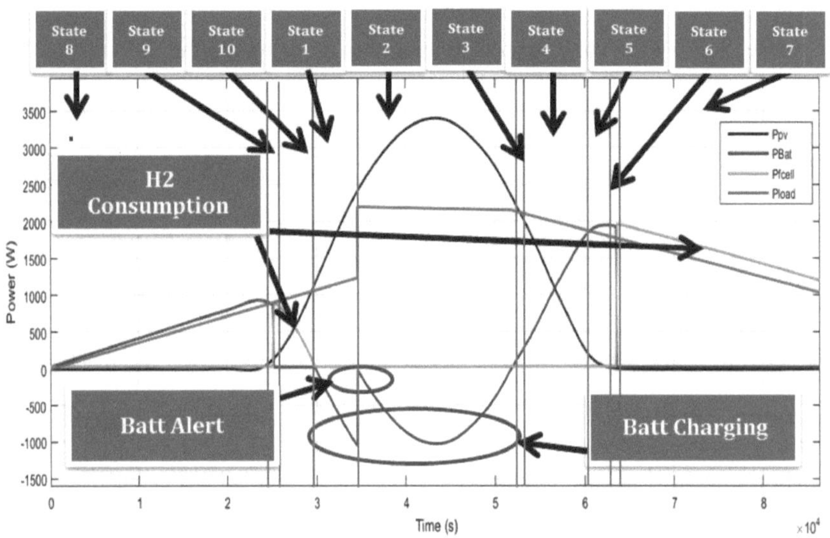

Fig 15: HES management control strategy

The proposed management strategy is essentially based on simple rules. Indeed, many management scenarios can be considered depending on the available energy sources, the energy consumption and the state of charge of the battery during 24 hours in order to ensure the correct operation according to a proposed operation algorithm. In this context, we used the Matlab/Simulink environment, which allows us to determine at each moment; the operation of the different elements that compose our hybrid energy system for climate change, as shown in Figure 14.

And for a change of random load defined as shown in the figure 15 which represents the variations of the powers of the elements of the HES to be studied:

5.6.HES power management scenarios (PV/PEMFC/Batteries)

Six scenarios can be discussed following figure 15 above and based on the experimental runs taken in figure 14 where the variation steps are as follows:

- From 00:00 a.m. to 08:00 a.m.: radiation (G = 0) and temperature (T = 2°C).
- From 08:00 a.m. to 12:00 p.m.: radiation and temperature increase rapidly.
- From 12:00 p.m. to 1:30 p.m.: the radiation is maximum (G = max) and the temperature is also maximum (T = max).
- From 1:30 p.m. to 6:00 p.m.: radiation (G) and temperature (T) begin to decrease.
- From 6:00 p.m. to 12:00 a.m.: at night, no radiation (G = 0) and the temperature tends towards (T = 3.5 ° C).

a) Scénarios1. (PV ON et PEMFC/Batteries OFF) **:** The PV generator is alone in active

state. The total power of the SEH is given by: $P_{ch} = P_{pv} + P_{fc} + P_{bat}$ with $P_{fc} = 0$ and $P_{bat} = 0$

-The batteries are initially half charged ***60% ≤ SOC≤ 98%.***

-The management system mainly takes care of the power supply for the load and the batteries.

This scenario can give three possible states:

State 1: The GPV alone which supplies the load and satisfies the SEH $\boldsymbol{P_{pv} \geq P_{ch}}$, in the case where $\boldsymbol{P_{pv} = P_{ch}}$ the battery remains in a state of alert for energy production (case of shade for example or case of bad conditions climate).

State 2: This is the case when the solar radiation reaches the maximum, the excess or the surplus of the GPV charges the batteries to ***SOC < 98%.***

State 3: The GPV powers the load only $\boldsymbol{P_{pv} = P_{ch}}$, no excess PV.

b) Scenarios2. (PV/Batteries ON and PEMFC OFF): The PV generator and the Battery are in Active state. The state of charge (SOC) of the battery is given by the following figure:

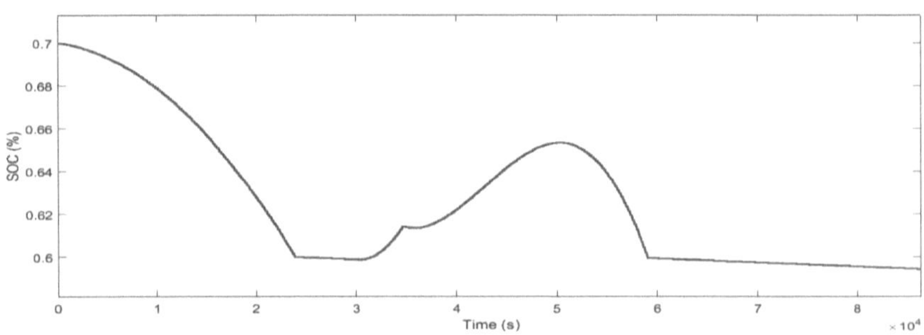

Fig 16: State of charge of the Li-Ion battery

State 4: The GPV is unable to supply the load $P_{pv}<P_{ch}$, the battery is needed because

$SOC > SOC_{min}$ to satisfy the SHE, so $P_{ch}= P_{pv} + P_{bat}$ with $P_{fc}=0$.

State 5: The battery alone which supplies the load because $SOC > SOC_{min}$ **or** $P_{bat} > P_{ch}$

Solar radiation begins to decrease with sunset or $P_{pv} = 0$.

State 6: The battery is begins to discharged $SOC < SOC_{min}$ and $P_{pv}=0$ or $P_{ch}= P_{fc} - P_{bat}$

The PEMFC takes the position to supply the load with a significant consumption of hydrogen.

c) Scénarios3. (PV/Batteries OFF and PEMFC ON) :

We can distinguish the following state:

State 7: The power P_{pv} is totally insufficient $P_{pv} = 0$, $P_{bat} = 0$ and $P_{fc} > P_{ch}$

The battery goes into reserve) the load immediately requests energy from the PEMFC.

In this case, the hydrogen consumption reaches its maximum point. The PEMFC ensuring the supply up to the point t=0 where the battery simultaneously takes over the total production.

d) Scenarios4. (PV/PEMFC OFF and Batteries ON): The PV generator and PEMFC

are in active state, $P_{pv} = 0$ *et* $P_{fc}=0$

State 8: only the battery supplying the load, $SOC \leq 98\%$ and $P_{bat} > P_{ch}$

e) Scenarios5. (PV OFF and Batteries /PEMFC ON) : The PV generator is in an idle state $\boldsymbol{P_{pv}=0}$ and $\boldsymbol{P_{ch}=P_{fc}-P_{bat}}$

State 9: The PEMFC begins to participate in supplying the load with the battery until the battery discharges $\boldsymbol{P_{bat}=0}$ and the solar radiation begins to increase with sunrise.

f) Scenarios6. (Batteries OFF and PEMFC/PV ON): $\boldsymbol{P_{ch}=P_{fc}+P_{pv}}$ avec $\boldsymbol{P_{bat}=0}$

State 10: The Battery discharges, it calls for help from the PEMFC hence a small consumption of hydrogen is necessary.

5.7. Control of Hydrogen

The variation of the hydrogen consumption of the PEMFC is similar to that of the current of the PEMFC (figures 17 and 18), which introduces that the production of electrical energy delivered by the PEMFC depends directly on the quantity of the molar flow rates of hydrogen consumed.

The hydrogen is produced by an electrolyser powered by excess energy from GPVs. When this energy is insufficient to supply the load, the previously stored hydrogen is used by the PEMFC to produce electricity. The aim of several studies is to minimize the consumption of hydrogen.

Finally, to achieve an efficient energy system, the key points are [39] [40] [41] [42] [43] [44]:

- The sizing of the components of the system.

- The electrical architecture and associated energy management.
- A suitable control strategy.
- The hydrogen storage system.

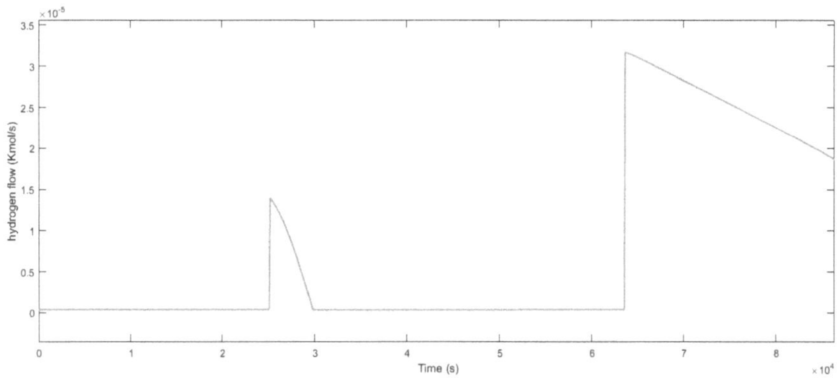

Fig 17: The flow of hydrogen consumed by the PEMFC

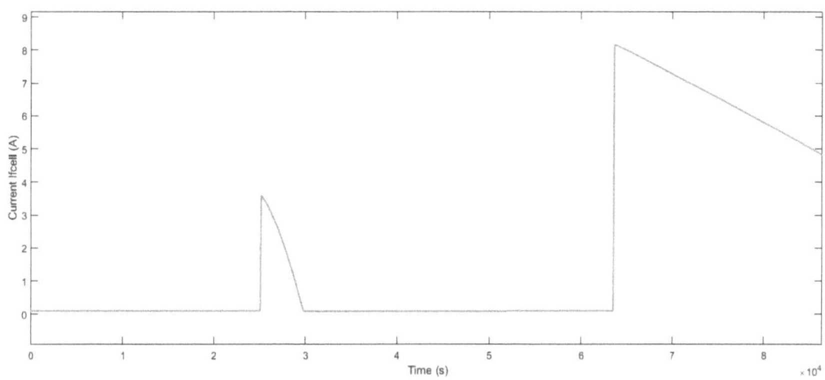

Fig 18: The Current produced by the PEMFC

An experimental study was made by [37][38], where he demonstrated that the storage volume depends only on the annual profile of the load and the gross energy, that is to say the energy supplied by the PVs and by the Fuel cell.

The energy supplied directly by the PV to the load (PV Direct) depends on the load profile.

By favouring the direct power supply, the gross energy is greatly increased; on the other hand, the daily consumption of hydrogen is reduced. The storage volume increases because the raw energy is more important [39] [40] [41] [42] [43] [44].

6. Conclusion

This research work is essentially based on a technical-economic feasibility study of off-grid hybrid energy systems, which is designed to provide electricity to remote regions and isolated urban regions in southern Algeria.

The hydrogen-based power generation technology as well as hybrid generation, hydrogen distribution and storage methods are the subject of continuous research and development as a huge a number of pilot projects currently underway in the field of renewable energy will be validated.

This paper shows the behaviour of an autonomous hybrid energy system which consists of a photovoltaic system, a battery system and a PEMFC fuel cell. Each production source is modelled individually, with its characteristics in the face of climate change.

A sizing was carried out after estimating the energy potential and the energy requested by the user.

An energy management strategy of the hybrid system studied has been developed and proposed with the macroscopic energy representation, which is a very all-powerful tool for synthetically describing and modelling complex multi-physical systems which use a simple inversion method, the maximum control structure which is designed to control each input / output element influencing directly on the adequate implementation of the overall hybrid system or on the components of this system. If this modelling formalism may seem close to the Bond-Graph, it differs from it in particular by making it possible to easily propose an MCS for the hybrid system considered and finally to deduce the appropriate practical control structure.

The results of the simulation of the entire hybrid system show the need for hybridization, especially in isolated sites, thus the effectiveness of this device in minimizing the consumption of hydrogen.

It is expected that in the near future, the various hydrogen fuel cell hybrid equipment, in general, will reduce costs, gaining a competitive advantage over other technologies in the field of fuel management and storage production energy.

References

[1] IPCC. Special Report: ''Global Warming 1.5°C - Summary for Policymakers, Tech. rep, Intergovernmental Panel on Climate Change'; 2018. *https://www.ipcc.ch/_sr15/chapter/spm/.*

[2] Kristov L, De Martini P, Taft JD. "A tale of two visions: designing a decentralized transactive electric system''. IEEE Power Energies Mag 2016Vol. 14 N° pp. 363–369. *https://doi. 2016. org/10.1109/MPE.2016.2524964. https://ieeexplore.ieee.org/document/* 7452738.

[3] Markard J. "The next phase of the energy transition and its implications for research and policy". Nat Energy 2018;3:628–633. *https://doi.org/10.1038/s41560-018-0171-7. https://www.nature.com/articles/s41560-018-0171-7.*

[4] Mancarella P. " MES (multi-energy systems): An overview of concepts and evaluation models". Energy 2014. Vol 65, pp. 1-17. *https://doi.org/10.1016/J.ENERGY.2013.10. 041. https://www.sciencedirect.com/science/article/pii/S0360544213008931.*

[5] Ivalin Petkova. Paolo Gabriellib. "Power-to-hydrogen as seasonal energy storage: an uncertainty analysis for optimal design of low-carbon multi-energy systems*". Applied Energy*. Vol 274, pp. 1-25. 2020.

[6] N. Benchouia, A.E. Hadjadj, "Modelling and validation of fuel cell PEMFC", *Journal of Renewables Energy*, Vol. 16 N°2, pp. 365-377, 2013.

[7] J. C. Amphlett, R. F. Mann, B. A. Peppley, P. R. Roberge, and A. Rodrigues, "A model predicting transient responses of proton exchange membrane fuel cells". *Journal Power Sources.* Vol. 61, pp. 183-188, 1996.

[8] L. Boulon, M-C. Pera, P. Delarue, A. Bouscayrol, and D. Hissel,"Causal Fuel Cell System Model Suitable for Transportation Simulation Applications", *ASME Journal of Fuel Cell Science and Technology*, Vol. 7, 2010.

[9] K. Mammar, A. Chaker, "Neural Network-Based Modeling of PEM fuel cell and Controller Synthesis of a stand-alone system for residential application", *International Journal of Computer Science,* pp.1694-0814, 2012.

[10] Gabrielli P, Gazzani M, Mazzotti M. "Electrochemical conversion technologies for optimal design of decentralized multi-energy systems: Modeling framework and technology assessment. *Applied Energy*. Vol 221, pp. 557-575. 2018.

[11] M. Götz, J. Lefebvre, F, and All. "Renewable Power-to-Gas: A technological and economic review". *Renewable Energy*. Vol 85, pp.1371-1390. 2016.

[12] F. Saadaoui. Contribution to the Multi-Physical Modeling and Control of a Hybrid Energy System (Photovoltaic, Fuel Cell and Batteries)". Tahri Mohamed University Bechar. Doctoral thesis. Defended June 11, 2020.

[13] L. Boulon, D. Hissel, A. Bouscayrol, and All, "From Modeling to Control of a PEM Fuel Cell Using Energetic Macroscopic Representation", *IEEE transactions on Industrial Electronics.* Vol 57, N° 6, pp. 1882-1891. JUNE 2010.

[14] A. Bouscayrol, B. Davat, B and All. "Multi-machine multi-converter system for drives: analysis of coupling by a global modeling, in Industry Applications", *Conference Record of the 2000 IEEE*. Vol. 3. pp. 1474-1481. 2000.

[15] L. Boulon. "Multiphysics modeling of energy storage and conversion elements for hybrid electric vehicles. Systemic approach to energy management". *Universite of Franche-Comte*, 2009.

[16] A. Bouirek. "MER-Backstepping hybrid design approach of correctors". *Doctoral thesis. University of Quebec Three Rivers. Canada*. August 2016.

[17] International School-REM16. Montreal, "Modeling and control based on Macroscopic Energy Representation: Applications to electric or hybrid vehicles and others". *University of Sherbrooke Campus of Longueuil, Quebec, Canada.* June 27th – 30th, 2016.

[18] J.G Carton, V. Lawlor, A.G Olabi, C. Hochenauer, G. Zauner, "Water droplet accumulation and motion in PEM (Proton Exchange Membrane fuel cell mini channels", *Journal of Energy*. Vol 39, pp 63-73, 2012.

[19] K. Agbossou, Y. Dube, N. Hassanaly, K. P. Adzakpa, and J. Ramousse. "Experimental Validation of a State Model for PEMFC Auxiliary Control, *IEEE trans. on Industry Applications*. Vol. 45, N°.6, pp. 2098-2103, 2009.

[20] S. Laribi, K. Mammar and all, "Impedance model for diagnosis of water management in fuel cells using artificial neural networks methodology", *International Journal of Hydrogen Energy*. Vol 41, No 38, pp. 17093-17101, 2016.

[21] F. Saadaoui. K. Mammar. A. Hazzab. "Water Distribution and the Impact of Relative Humidity in a PEMFC Energy System using Macroscopic Energy Representation by Inversion Control". *Majlesi Journal of Electrical Engineering*. Vol. 15, No. 3, September 2021.

[22] K. Mammar, F. Saadaoui and S. Laribi. "Design of fuel cell model for flooding and drying diagnosis using fuzzy logic clustering". *Renewable Energy Focus*. Vol. 30. pp. 123-130. September 2019.

[23] F. Saadaoui. K. Mammar. A. Hazzab. "Modeling of photovoltaic system with maximum power point tracking control by neural networks". *International Journal of Power Electronics and Drive System (IJPEDS)*. Vol. 10. No. 3, pp. 1575-1591. September 2019.

[24] J.S. Kumari, Dr.Ch. Sai Babu and A. Kamalakar Babu. "Design and Analysis of P&O and IP&O MPPT Techniques for Photovoltaic System". *International Journal of Modern Engineering Research (IJMER)*. Vol. 2. N°. 4, pp. 2174-2180. 2012.

[25] Gh. Remy, O. Bethoux , C. Marchand, and H. Dogan. "Review of MPPT Techniques for Photovoltaic Systems". *Paris Electrical Engineering Laboratory (LGEP) / SPEE-Labs, CNRS UMR 8507, SUPELEC; Pierre and Marie Curie University P6; Paris-Sud University 11, street Joliot Curie, Plateau of Moulon F91192 Gif sur Yvette CEDEX.* France, pp. 1-14.

[26] K.S. Akbli. "Multiphysical modeling of the energy flows of a Photovoltaic-Electrolyzer PEM-Fuel Cell PEM coupling for a Stationary application". *Cocody University Abidjan. Physics Doctoral Thesis.* Defended March 6, 2012.

[27] F. Antoine Lebel. "Sizing method and modeling of lithium-ion batteries". *University of Sherbrooke Quebec Canada. Doctoral thesis.* Defended on April 28, 2017.

[28] An LI. "Experimental analysis and modeling of battery cells and their assemblies – Application to electric and hybrid vehicles". *University of Claude Bernard Lyon 1. Doctoral thesis*. Defended on April 23, 2013.

[29] Mr Rebhi. "Technical-economic study of a photovoltaic-diesel hybrid system in a Saharan environment". *Tahri Mohamed University Bechar. Doctoral thesis*. Defended 2018.

[30] L. Zrourou and All, "Dimensioning of a solar installation for an isolated house in Béjaia*", International Conference ICRE'08*, Ghardaia, Algeria, 2008.

[31] Akassewa Tchapo SINGO. "Photovoltaic power system with hybrid storage for energy-autonomous housing". *Henri Poincaré University*, Nancy-I, 2010.

[32] L. Boulon, M-C. Pera, P. Delarue, A. Bouscayrol, and D. Hissel. "Causal Fuel Cell System Model Suitable for Transportation Simulation Applications", *ASME Journal of Fuel Cell Science and Technolog.* Vol. 7, 2010.

[33] K. Mammar, A. Chaker, "Neural Network-Based Modeling of PEM fuel cell and Controller Synthesis of a stand-alone system for residential application*", International Journal of Computer Science.* pp.1694-0814, 2012.

[34] L. A. M. Riascos. "Relative humidity control in polymer electrolyte membrane fuel cells without extra humidification". *Journal of Power Sources*. Vol. 184, pp. 204-211, 2008.

[35] Chrysovalantou Ziogou and All. "Infrastructure, automation and model-based operation strategy in a stand-alone hydrolytic solar-hydrogen production unit". *International Journal of Hydrogen Energy*. Vol. 37, pp. 16591-16603. 2012.

[36] Manuel Castan˜eda and All. "Sizing optimization, dynamic modeling and energy management strategies of a stand-alone

PV/hydrogen/battery-based hybrid system". *International Journal of Hydrogen Energy. Article in press*. Vol. XXX, pp. 1-16. 2013.

[37] Mr. Zandi. "Contribution to the management of hybrid sources of electrical energy". *Nancy University - National Polytechnic Institute of Lorraine. Doctoral thesis*, 2010.

[38] S. Busquet. " Study of an autonomous energy production system coupling a photovoltaic field, an electrolyser and a fuel cell: creation of a test bench and modelling. *Ecole of Mines, Paris. Doctoral thesis.* Defended on December 15, 2003.

[39] Samson Gebre, Tore M. Undeland(IEEE Fellow) and All. "Optimal Load Sharing Strategy in a Hybrid Power System based on PV/Fuel Cell/ Battery/Supercapacitor". *2009 IEEE.* 978-1-4244-2544-0/08/$20.00 ©2009.

[40] M.M. Samy, S. Barakat, H.S. Ramadan. "A flower pollination optimization algorithm for an off-grid PV-Fuel cell hybrid renewable system". *International Journal of Hydrogen Energy.* Vol. 44, pp. 2141-2152. 2019.

[41] Raluca-Andreea Felseghi and All. "Optimal Synergy between Photovoltaic Panels and Hydrogen Fuel Cells for Green Power Supply of a Green Building-A Case Study". *Sustainability Journal.* 13, 6304. https://doi.org/10.3390/su13116304, pp. 1-20. 2021.

[42] Premananda Pany. "Hybrid PV Solar, PEM Fuel Cell and Battery with Active Power Management for Electric Vehicles". *International Journal of Engineering and Advanced Technology (IJEAT)* ISSN: 2249 – 8958, Vol.9 Issue-1, pp. 2017-2020. October 2019.

[43] Vipin Das and All. "Optimal Sizing and Control of Solar PV-PEMFC Hybrid Power Systems". *International Journal of Mathematical, Engineering and Management Sciences*. Vol. 6, No. 4, pp. 1137-1156, 2021.

[44] Salsabil Gherairi. "Design and implementation of an intelligent energy management system for smart home utilizing a multi-agent system". *Ain Shams Engineering Journal* xxx (xxxx) xxx. Article in press 2022.

TABLE OF CONTENT

Kolibri 360 GmbH
Ferdinand-Jühlke-Straße 7
99095 Erfurt, Deutschland
produktsicherheit@kolibri360.de

Druck:
CPI Druckdienstleistungen GmbH
im Auftrag der
Kolibri 360 GmbH
Ferdinand-Jühlke-Str. 7
99095 Erfurt